YOUR KNOWLEDGE HAS VALUE

Bibliographic information published by the German National Library:

The German National Library lists this publication in the National Bibliography; detailed bibliographic data are available on the Internet at http://dnb.dnb.de .

Imprint:

Copyright © 2015 GRIN Verlag, Open Publishing GmbH
Print and binding: Books on Demand GmbH, Norderstedt Germany
ISBN: 978-3-668-07508-5

This book at GRIN:

http://www.grin.com/en/e-book/308947/got-milk-the-consumption-of-milk-and-what-it-does-to-our-body

Alex Monseur

Got Milk? The consumption of milk and what it does to our body

GRIN Publishing

GRIN - Your knowledge has value

Since its foundation in 1998, GRIN has specialized in publishing academic texts by students, college teachers and other academics as e-book and printed book. The website www.grin.com is an ideal platform for presenting term papers, final papers, scientific essays, dissertations and specialist books.

Visit us on the internet:

http://www.grin.com/

http://www.facebook.com/grincom

http://www.twitter.com/grin_com

Alex Monseur

11 June 2015

English 112

Got Milk?

Human beings should not be drinking milk past infancy from any other species because it is unhealthy, unethical, and unnatural. Many don't question or even think about what they consume on a day-to-day basis because it's what's familiar. Many don't want to know because change or knowing the truth can be uncomfortable. The western diet is by far one of the worst diets in the world and dairy is one of the worst things you can consume (Woolston.) Going to the grocery market, dairy products paint a picture of cows being these happy, frolicking, farm animals. But truthfully, cows endure immense suffering on factory farms. In fact, the dairy industry has made the life of a cow one of the most painful lives to live. In America, cows are exploited and taken advantage of everyday, and largely for the taste of their breast milk. Cows go through a strenuous amount of torture and as many don't believe it, are sentient beings. In commercials, breakfast is typically shown accessorized with a large glass of milk. The United States food pyramid even includes dairy. The popular advertisements "Got Milk?" have been encouraging milk consumption since 1993 (Got Milk.) The companies' slogan is "Drink to a brighter future" (Got Milk.) But what if the future isn't bright? What if consuming milk products actually dims your future and makes you more susceptible to developing hazardous health problems?

Contrary to what most of society has been brainwashed to believe, consuming milk products is very detrimental to our health. Milk contains saturated fat and cholesterol, which can lead to a number of chronic diseases, such as heart disease

1

(Milk Myths.) Cow's milk is also an acid forming when consumed. This causes an acidic environment in the body, which illness like, cancer, heart disease, and bacteria thrive off of (Campbell.) "The scientist, Ganmaa Davaasambuu, M.D., Ph.D., a native Mongolian, noted that ingestion of natural estrogens from cows (particularly from pregnant cows) in milk may be linked to breast, prostate, and testicular cancers in humans" (Weil.)

Jane Plant, a geochemist, was diagnosed with breast cancer in 1987, and in 1993, developed a tumor the size of an egg in her neck (Hicks.) The doctors told her she would not live for more than a couple months (Hicks.) Professor Plant and her husband have both worked in China on environmental issues and knew Chinese women had a very low rate of breast cancer, if any (Hicks.) Her and her husband took time to analyze why and figured it was because the Chinese do not consume any dairy (Hicks.) Professor Plant had nothing to lose so she switched to a dairy-free Asian-styled diet and within 6 weeks, the lump in her neck was gone and she's lived in remission for 18 years (Hicks.)

Milk today is more harmful than it's ever been and is highly processed and contains harmful chemicals, one in particular, a growth hormone called "Insulin" (Rietz.) This growth hormone is considered to be a "'fuel cell' for any cancer" (Rietz.) Once a child grows out of infancy, the body slows the production of "lactase" which is an enzyme that allows mammals to digest the lactose in milk (Phelan.) Without the enzyme, the lactose will rot in the gut (Phelan.) Regardless of what people have been trained to believe, human beings do not need a drop of milk after weaning (Rietz.) ""I no longer recommend dairy products after the age of 2 years. Other calcium sources offer many advantages that dairy products do not have" said by

Dr. Benjamin Spock (Rietz.) Just to name a few alternative calcium sources, there is: broccoli, oranges, oatmeal, almonds, kale, etc. (McClees.)

Besides dairy being completely harmful to human bodies, the process of getting dairy products is completely unethical. Cows are manipulated and exploited on a daily basis for human's selfish consumption. More than 9.3 million cows are continually raped and are artificially inseminated throughout the year using long mental probes (Milk Myths.) This is how the lactation process is started. Once the calf is born, it is only allowed a limited amount of time to feed until it is taken away from its mother and shackled inside a crate. If it is a male calf, then it is used for veal production (Milk Myths.) "Calves can become so distressed from separation that they become sick, lose weight from not eating, and cry so much that their throats become raw" (Milk Myths.) During this time, cows weep for their baby the same way a human would if their baby was taken from them (Milk Myths.)

"The very saddest sound in all my memory was burned into my awareness at age five on my uncle's dairy farm in Wisconsin. A cow had given birth to a beautiful male calf...On the second day after birth, my uncle took the calf from the mother and placed him in the veal pen in the barn—only ten yards away, in plain view of his mother. The mother cow could see her infant, smell him, hear him, but could not touch him, comfort him, or nurse him. The heartrending bellows that she poured forth—minute after minute, hour after hour, for five long days—were excruciating to listen to. They are the most poignant and painful auditory memories I carry in my brain." said Michael Klaper, M.D (Milk Myths.)

Cows are then hooked up to milk machines and are expected to produce unnaturally high levels of milk production (Milk Myths.) Every time one buys any milk product, they are contributing to this constant cycle of hell. And to make it worse, after 5 years of the same cycle, cows become exhausted, and are deemed profitless and useless and sent to the slaughterhouse (Milk Myths.)

It does not take much to think about that consuming cow's milk is completely unnatural. If one is not a cow, then one should not drink cow's breast milk. Human beings are the only species that drinks another species breast milk. One would probably not consider drinking any other mammals breast milk, so why is it that humans drink cows milk? Most humans have never questioned this because again, it's what we've grown up to see familiar. "It's not natural for humans to drink cow's milk. Human's milk is for humans. Cow's milk is for calves. You have no more need of cow's milk than you do rat's milk, horses milk or elephant's milk. Cow's milk is a high fat fluid exclusively designed to turn a 65 lb. baby calf into a 400 lb. cow. That's what cow's milk is for!" said Dr. Michael Klaper (Rietz.)

It is argued that humans have been drinking cow's milk for centuries, why stop now? First off, yes milk has been drank for centuries, but milk is now unhealthier than ever (Mercola.) Milk is now highly processed and has a lot of added harmful chemicals, such as hormones and antibiotics (Mercola.) And, just because something has been going on for years does not make it the right thing to do. One could argue that way about anything: murder, slavery, rape, etc.

It has also been argued that milk is really healthy for us because it supports our bones and contains an abundance of calcium. This myth needs to be debunked. Even though cow's milk contains calcium, human beings barely absorb the calcium in the milk, especially if it's pasteurized (Goldschmit.) Ironically, it actually increases calcium loss from the bones (Goldschmit.) "Cow's milk is a foreign substance that has pervaded every corner of our diets... Today there is little doubt that early and frequent feeding of dairy products leads to greatly increased incidence of childhood diabetes. It has been confirmed that high cow's milk consumption is a major cause of osteoporosis" (Palmer.)

This information may have been more detrimental to ones diet 10 years ago. But today, there is an abundance of milk alternatives that will have never have one missing milk again. To name them, there's: soymilk, almond milk, hemp milk, rice milk, coconut milk, cashew milk, the list goes on. It may be hard cutting every milk product out of one's diet but simply switching to a healthier and cruelty-free option to ones best ability could make a big difference.

Cow's milk is unhealthy for humans to digest, unethical considering the tortuous process that is involved, and unnatural considering we are the only species that drinks another species breast milk. So why is the majority of American society misinformed when it comes to the topic of milk? Why is milk made out to be this necessary thing we need in the American diet? The dairy industry is worth billions of dollars and there is absolutely no way they could ever risk losing their money. They need all the encouragement and advertisements they can get to keep their business thriving. It's up to the people to educate themselves and open their minds to see the hard truth about consuming cow's milk.

Works Cited

- Campbell, Meg. "Is Milk Alkaline or Acid?" *LIVESTRONG.COM.* LIVESTRONG.COM, 13 Mar. 2014. Web. 11 June 2015.

- Goldschmit, Vivian. "Debunking The Milk Myth: Why Milk Is Bad For You And Your Bones." *Debunking The Milk Myth: Why Milk Is Bad For You And Your Bones.* N.p., n.d. Web. 11 June 2015.

- "Got Milk? Drink to a Brighter Future." *Got Milk? Drink to a Brighter Future.* N.p., n.d. Web. 11 June 2015.

- Hicks, Cherrill. "'Give up Dairy Products to Beat Cancer'." *The Telegraph.* Telegraph Media Group, 2 June 2014. Web. 11 June 2015.

- MClees, Heather. "The Importance of Calcium and How to Get Enough Without Dairy." *One Green Planet.* N.p., 22 Sept. 2014. Web. 15 June 2015.

- Mercola, Dr. "A Glass of Milk Could Contain as Many as 20 Chemicals." *Mercola.com.* N.p., 26 July 2011. Web. 11 June 2015.

- "Milk Myths - Woodstock Farm Animal Sanctuary Woodstock Farm Animal Sanctuary." *Milk Myths - Woodstock Farm Animal Sanctuary Woodstock Farm Animal Sanctuary.* N.p., n.d. Web. 11 June 2015.

- Palmer, Linda, DC. "Linda Folden Palmer, DC - Milk - ProCon.org." *ProConorg Headlines.* N.p., 10 Jan. 2008. Web. 11 June 2015.

- Phelan, Benjamin. "The Mysterious, Mutant, Civilizing Power of Milk." *Evolution of Lactose Intolerance.* N.p., 23 Oct. 2012. Web. 15 June 2015.

- Rietz, Dave. "The Truth About Milk - Read." *The Truth About Milk - Read.* N.p., 24 June 2002. Web. 11 June 2015.

- Weil, Andrew, Dr. "Q & A Library." *Does Milk Cause Cancer?* N.p., 30 Mar. 2007. Web. 11 June 2015.

- Woolston, Chris, M.S. "What's Wrong With the American Diet?" *What's Wrong With the American Diet?* N.p., 15 Mar. 2015. Web. 15 June 2015.

YOUR KNOWLEDGE HAS VALUE

- We will publish your bachelor's and
  master's thesis, essays and papers

- Your own eBook and book -
  sold worldwide in all relevant shops

- Earn money with each sale

Upload your text at www.GRIN.com
and publish for free